Pedro Barroso
Joseane Farias Souza

Digital Archival Repositories

Pedro Barroso
Joseane Farias Souza

Digital Archival Repositories

from trust to preservation

ScienciaScripts

Imprint

Any brand names and product names mentioned in this book are subject to trademark, brand or patent protection and are trademarks or registered trademarks of their respective holders. The use of brand names, product names, common names, trade names, product descriptions etc. even without a particular marking in this work is in no way to be construed to mean that such names may be regarded as unrestricted in respect of trademark and brand protection legislation and could thus be used by anyone.

Cover image: www.ingimage.com

This book is a translation from the original published under ISBN 978-620-2-17211-0.

Publisher:
Sciencia Scripts
is a trademark of
Dodo Books Indian Ocean Ltd. and OmniScriptum S.R.L publishing group

120 High Road, East Finchley, London, N2 9ED, United Kingdom
Str. Armeneasca 28/1, office 1, Chisinau MD-2012, Republic of Moldova, Europe
Printed at: see last page
ISBN: 978-620-7-19530-5

Index :

UEPB

PARAÍBA STATE UNIVERSITY-UEPB
CENTER FOR APPLIED SOCIAL AND BIOLOGICAL SCIENCES-CCBSA
CAMPUS V MINISTRO ALCIDES CARNEIRO
BACHELOR'S DEGREE COURSE IN ARCHIVOLOGY

PEDRO AUGUSTO DE LIMA BARROSO

DIGITAL REPOSITORIES: HISTORY AND CHARACTERISTICS

JOÁO PESSOA
2017

To my parents, for their dedication, education, encouragement and guidance, I DEDICATE this Final Paper.

ACKNOWLEDGMENTS

Many people are part of the construction of this research. I would like to thank everyone who has contributed in some way, so that I can carry out each stage of this research until it reaches its culmination, which is the day it is presented.

Despite thanking everyone far too much, we have those people who require special affection, because they are part of our daily lives, they support us and make us reach our main goal.

To my parents, my brothers and my sister-in-law in particular, who are always with me, and are the ones who here on earth fulfill all my wishes and I know that in difficult times they will be with me.

To my friends, after the exchange I was able to realize how big they all are, but in particular I would like to thank five of them who have always been with me during these years of my degree: Thamires Neves, Neto Chaves, Valter Filho, Waldemar Lucas, Lucas Santos and Jurandir Lourenço and to all of them my sincere thanks, because without you, I could not have completed another phase of my life.

And most profoundly, thank you to Joseane Farias and Katya for everything they did for me during my degree.

To my dear teachers from undergrad, especially Josemar Henrique de Melo and Eliete Correia dos Santos, who instilled great confidence in me and always insisted that I never give up on my goals. And corroborating these factors academically, I also have a lot to thank my friend Thais for all her zeal and dedication in all the mentoring she did in 2017, as well as everything Professor Manuela Pinto did for me during my internship abroad at the University of Porto.

During the exchange period in the Archivology undergraduate program, I created bonds that are forever marked and these bonds directly participated in my sleepless nights doing this work in the city of Porto, and the three people to whom I owe a lot of strength and perseverance: Eric Bortoleto (my brother), Maira Gondim and Rafael Cabral those who are my daily nostalgia and I owe my personal and professional maturation to you.

"The introduction of the preservation of digital archival documents is not only a new challenge, but also of the utmost importance for Brazilian archives, since the preservation of these documents will guarantee the memory, history and rights of future generations, given that the current trend is to produce more and more digital documents" (INNARELLI, 2011, p. 34).

SUMMARY

Technological evolution has led to a posture of innovation. When we refer to information processing, this is especially true in the digital age, because every day we are faced with a large amount of information in the most varied contexts, thus leading us to think about how information is being preserved so that it can be consulted in the future. The general aim of this research is to show the history of digital repositories and their characteristics. As for the methodology used, this is a bibliographical, exploratory study with a qualitative approach. Therefore, digital repositories have emerged with the main aim of making information open to the entire community, since the beginning of scientific communication at the end of the Second World War, and we realize within the research that we have been able to detect this entire path, highlighting that this study is only the beginning of a research, which due to technological evolution is expanding and causing new researchers to emerge on the subject.

Keywords: Archivology. Digital Repository. Information Preservation. Information Management.

Chapter 1
1 INTRODUCTION

Technological evolution has led to a posture of innovation. When we refer to the processing of information[1] , it is mainly in the digital age, because every day we are faced with a large amount of information[2] in the most varied contexts, thus leading us to think about how information is being preserved so that it can be consulted in the future.

The processes and actions involving documents, especially those related to the organizational context, have been following this technological evolution in terms of storage, search methods, processing and preservation over a long period of time. In this context, the creation of measures to guarantee the preservation of the original context of documents is gaining prominence within institutions.

With the challenges posed by the storage of information produced in a digital environment, a tool has emerged that facilitates the permanent preservation of digital documents: Digital Repositories.

From the perspective of Santos et al. (2009), the first repositories arose from the need that human beings had to store and access information from information units (archives, libraries and museums).

Barreto (2010) emphasizes that repositories, in the context of information units, emerged in the context of universities and were joined by the Open Access movement. However, even with this origin, in the context of institutional production, they are used for the following activities: archiving, disseminating and preserving other types of originals and documents, such as articles, among others.

In view of this scenario, this research is concerned with how this whole process of emergence was carried out, with the following problem: What is the history of digital repositories and their characteristics?

With this in mind, the general aim of this research is to show the history of digital repositories and their characteristics, bringing this context into relation with Archivology and highlighting some of the software used in Librarianship.

In order to achieve the general objective of the research, objectives -

[1] The process that information goes through until it reaches the user.

[2] According to Araújo (2014), the explosion of information occurred at the end of the Second World War, as the results of scientific activities and technological advances manifested themselves as war strategies and due concern began to be shown for the information produced in the scientific context.

which are constructed to facilitate understanding of the subject:

- Define what a digital repository is;
- Characterize some of the types of repositories;
- Present the characteristics of Digital Archival Documents-DAD;

- Address the *Open Archive Information System (*OAIS) model;
- Exhibiting digital repositories in the field of Archivology and Librarianship.

Therefore, all this contextualization within the research is justified, because in order to have knowledge on the subject it is important to know every detail of this tool, which since its emergence has made access to information even easier.

This work was carried out during the student's period of academic mobility at the University of Porto, where he had the opportunity to take the Information Preservation course taught at the time by Professor Maria Manuela Pinto, a partner in the Archivology course at the State University of Paraíba (UEPB), who, when taking the course, sparked research into the area, and so the research began during the exchange period and was finalized on his return to the University of origin, as the subject of his Course Conclusion Paper (TCC).

1.2 Methodology

This is a bibliographical study of a basic nature, in which we used exploratory research in the scope of the research, in which several authors emphasize this type of investigation, corroborating the advancement of the theme.

According to Vergara (1998, p.45), exploratory research occurs when the knowledge being studied is still little covered in the field of knowledge and is interconnected with the survey that is worked on during the investigation of our object of study and together with this the descriptive objective is used, which is worked on to describe the general characteristics of digital repositories together with their history.

As for the nature of the research, it can be classified as qualitative, because this type of approach requires us to describe qualities.

> In the qualitative approach, the scientist aims to deepen her understanding of the phenomena she is studying - the actions of individuals, groups or organizations in their environment or social context - by interpreting them from the perspective of the subjects themselves who participate in the situation, without worrying about numerical representativeness, statistical generalizations or linear accounts of cause and effect. Thus, we have the following fundamental elements in a research process (GUERRA, 2014, p.11).

It should be noted that each phase covered in this section was carried out in a succinct manner so that we could analyze the object of study with phases that had already been established and were directly related to the objectives.

1.3 Research structure

This work is divided into six chapters. In the first chapter, we show the contextualization of our object of study, i.e. everything that goes back to its introduction, justification, research objectives and the factors that guided it.

After the initial contextualization, the second chapter emphasizes all the theoretical foundations related to the research, until reaching the section on Digital Archival Documents (DAD), which facilitates understanding about preservation, security and what metadata is.

In the third chapter, we go into the subject of our object of study, digital repositories, where we look at their history, their relationship with digital archives, the *Open Archive Information System* (OAIS) conceptual model.

In the fourth chapter, the types of repositories that exist are emphasized, as are the platforms that facilitate their use and, finally, their relationship and applicability with *software* in the field of archives is shown and two technological devices used in librarianship are emphasized.

Continuing the work, in the fifth chapter, the existing security parallels within the software are parameterized, in other words, what is necessary to be able to have secure information within digital repositories.In the last chapter, all the final considerations and suggestions regarding the theme are constructed.

Chapter 2

2 THEORETICAL BACKGROUND

As an essential field of study for Archivology, information has become complex due to the undeniable advance of information, which has led to the emergence of new guidelines and the broadening of its field of study, since it is the object of Archivology.

Due to the complexity surrounding the word information, in this section we will present speeches that will make it easier to understand the concept of information in the archival scenario, which has been called "archival information" since its beginnings, and the themes that make up the information systems in which they are stored and consulted on digital platforms in order to streamline access for users.

Since the emergence of archivology, information has been characterized as a source of evidence framed in the body of the document, i.e. recorded on a medium, be it physical or digital, which is why the term archival information was coined.

Following the perspective of Paes (2007), in the process of emergence documents were used to establish or claim rights, and when their demands ended they were transferred to archives or document depositories[3] as archives were called in the early days.

In Archivology, information is not presented as a simple aspect, however, it takes on a new format and concept, coming to be understood as archival information, according to Conceigao (2013) archival information is defined as that which is generated and received and which is managed by companies or people during the period in which they carry out their activities, Its aim is to satisfy information needs, regardless of the form used, enabling management to carry out its duties and functions quickly, efficiently and economically, as well as seeking to safeguard the rights and duties attributed to documents.

It can be said that information is the set of representations and dialogues, which depends not only on the medium, but also on how it is understood. From production to access, we will refer below to an image that represents this in Figure 1.

Figure 1 - Processing the Informando

---

[3] According to Paes (2006), the first evolution of the word *arché* was *archeion,* which followed the same meaning as the word *archeion,* which is the place where documents are kept and deposited.

Source: Adapted from Pinto (2016).

When we look at the information process in figure 1, we can see that it represents the phases that archival information goes through until it reaches its apex, which is its dissemination to the user, and we can also point out that this process does not currently adopt a sequence, According to Pinto (2015), the management carried out during the process does not follow the linearity of a management cycle, but requires the existence of support services that are called upon not in sequence, but whenever necessary, with the information production phase acquiring increasing importance.

After the processing phases have been completed, the archival information is added to the information system, which contextualizes it so that it can be made available and consulted. In turn, the concept of Information System (IS), a singular term with several related concepts, is defined in the Dictionary of Terminology in Information Science (DELTCI, 2017) as being made up of the different types of information recorded or stored externally to the subject (what each person has in their memory is information from the system), regardless of the support (material or technological), according to a structure (producing/receiving entity) extended over time.

In organizations, the structure that supports the system is of great importance, reflected in the information produced and accumulated and being captured, for example, through the technique of organic functional analysis and direct, participant or retrospective observation, as defined in the technical pole of the quadripolar methodological approach[4].

The IS cannot be confused with the information technology system (ITS), which is different but inseparable. The ITS corresponds to the technological platform - the physical and logical environment or medium - that supports the production, processing, circulation, storage, transmission and access to information that constitutes the IS itself. According to the perspectives analyzed, these systems have a whole set of parameters to build on, since their main characteristic is to facilitate the

---

[4] According to Guimaraes et.al (2013), the four-pole approach consists of four poles: epistemological, theoretical, morphological and technical. The advantage of the four-pole method stems from the post-positivist, systemic and constructivist thinking that underlies it.

flow of information (regardless of the medium), but taking into account the information needs of companies and individuals.

> [...] an Information System is a totality formed by the dynamic interaction of its parts, in other words, it has a lasting structure with a flow of states over time. Therefore, an Information System is made up of the different types of information recorded or not externally to the subject (what each person has in their memory is system information), no matter what the support (material or technological), according to a structure (producing/receiving entity) prolonged by action on the timeline (DELTCI, 2017, *Online)*.

And in a more complete way, digital platforms have emerged, with the same aim of disseminating and making information available. Silva (2012) reports that a digital platform is a technological base with the principle of storing, retrieving, disseminating, communicating and transforming the flow of information.

Silva (2012) indicates the following concept:

> I propose that we understand the digital platform as the 'space of inscription and transmission' of human and social information [...]. It is a technological 'space' which, in essence, is still made up of software [...] and hardware [...] in which various technologies and services converge in order to make it an instrument of infocommunicational mediation (SILVA, 2012, p.7).

Therefore, it is important to characterize this whole process in which archival information permeates, because in order to be able to dissertate on the contents of the following sections, it was necessary to explain this whole process, so that readers could understand the whole dimension added to archival information.

2.1 Digital archival documents: preservation, security and metadata

The intense process that takes place in the midst of the digital age, as presented above, leads us to consider all aspects of obtaining information until it is discarded. Within these phases, we equate the preservation and security of information with Information Management (IM), because there is concern about how these information can be kept intact and preserved for users.

And this fear can be emphasized in the following way: when information is stored within the system, it must be treated in an archival and technological manner, because both are interconnected. The integrity mentioned above refers to the information being stored in the same way from the moment it is produced until it is accessed by the user and preserved in accordance with the criteria that both professionals have chosen to apply to it, avoiding technological obsolescence,

With these concerns emphasized above, we can point out that the preservation of information on digital media deserves greater attention, since it can be used from the production of the Digital Archival Document (DAD) to its storage in the desired location.

> A digital archival document is a unit coded in binary digits and endowed with specific characteristics, which is created, transferred, processed and stored and/or disposed of using computer systems in their *hardware* and software structure (SOUZA et al, 2017, 291).

With this in mind, we have to determine what requirements must be met in order to carry out the due process, and the work of the archivist to direct these parameters begins.

Taking into account the perspective of Professor Flores and Schafer (2013), we can see that the DAD has a direct challenge with the preservation of information, because there are obstacles with information technology that are adjectivized as "innovative, agile elements, susceptible to constant and sudden changes, predominantly concerned with innovation [...]" (FLORES; SCHAFER, 2014, p.180).

Therefore, we need to parameterize a document that is ready for any change regardless of the technology applied, in other words, a DAD that is ready for possible transformations[5] , and when we talk about all this application we are referring to the *software* that has been and is being built from these perspectives, which are the digital repositories.

With this in mind, the DAD's mission is to avoid interference that alters the original format of the document, and to preserve all the information that is inserted during processing, as it represents an information need that has been solved or is still in the process of being solved.

Another important aspect is the security of this document in the digital environment, where Flores and Sfreedo (2012) take the view that the aim of this security parameter is to make information accessible in the right way, with all the appropriate security requirements, including authenticity. In the same research, they point out that there are guarantees to be followed, such as confidentiality, integrity and availability.

Another parameter used in DAD is metadata, which is fundamental in digital documents. It consists of taking into account the formats adopted by the *hardware* and *software* to arrive at their types: descriptive, structural, administrative and technical.

For Pinto (2016), metadata is "[...] information added automatically or semi-automatically to the document, at various points in its life cycle, and which will allow it to be contextualized under the different aspects already mentioned" (PINTO, 2016, slide 87).

Following the same idea about metadata, it is good to know that it is divided into six layers: context, management, terms and conditions, responsibility, usage history and structures. And another relevant aspect in this context is what we mentioned earlier about the precepts for information to be

---

[5] These transformations are related to technological obsolescence, which can be avoided by: encapsulation, emulation and migration used to avoid technological obsolescence.

considered trustworthy: : authenticity, trustworthiness, integrity and intelligibility.

Document description in physical format and in the digital sphere gave rise to descriptive metadata and within DAD, it involves, according to Pinto (2016), cataloguing records, search aids, specialized indexes, hyperlink relationships between resources and user notes, all based on the context in which the information is produced.

So this type aggregates the main information about the documents in digital form and is fed in as they are processed. All of this helps the information to be retrieved in the right way, and also to manage the digital documents. These factors help users to access content quickly and easily. One example of descriptive metadata software we can cite is *Dublin Core,* which was published in conjunction with the ISO standard and consists of the information in Table 1.

Table 1 - Types of Meta-informing[6] structured

Name	Funnao
Dublin Core Metadata Element Set (DOMES)	A set of 15 meta-information elements whose purpose is to facilitate the discovery of electronic resources. ISO, ANSI/NISO, CEN standard; Official position within the W3C.
Dublin Core Qualifiers (DCQ)	Element refinement - specifies the meaning of an element in more detail; Coding scheme - a scheme that helps interpret the value vocabularies.

Source: Pinto (2016, slide 160).

To finalize the issue of Dublin-core, we should show how it works at the description level, which consists of

> Records made from *Dublin Core* can describe "items" or a "collection". Although a detailed bibliographic description is only allowed for a level, the "relation" element ensures attributes such as "is part of" and "is part of" that make reference to the title (RAMALHO; VEIGA, 2007, p. 25).

It deals with structured metadata that lists the parameters for describing the internal structure of the document, both physical and digital.

The structured *metadata* that we are going to discuss is the *Metadata Encoding and Transmission Standard* (METS), according to Ramalho and Veiga (2007), which is a type of structured information that emerged in 2001 by the *Digital Library Federation,* and has seven main sections that show the interconnection of different structures following the direction of unification, they are presented in table 2:[6]

Table 2 - Brief description of the METS structure shows the interconnection of different structures moving towards unification

fvStructure	Function
Cabegalho	The header containing metadata describing the METS document itself, including creator, editor, etc.
Descriptive Metadata	This section can reference metadata external to the METS document (such as a UNIMARK or EAD record accessible on a server on the Internet), contain embedded metadata, or both. Multiple instances of metadata, internal or external, can be included in this section.
Administrative Metadata	This section provides information about the files that were created and stored, intellectual property rights, metadata about the original object from which the object was derived, etc.
Files section	This section lists all the files that make up the object. <fle> elements can be grouped into <fleGRP> file group elements.
Structural Map	This map is the heart of the METS document. It outlines a hierarchical structure for the object, and links the elements of this structure to files with content and metadata relating to each element.
Structural connections	This section makes it possible to record references between nodes in the hierarchy outlined in the Structure Map. This section is particularly valuable when using METS to archive websites.

Source: Ramalho and Veiga (2007, p. 27).

Ramalho and Veiga (2007) emphasize that, following on from the *Open Archive Information System (*OAIS) and METS, a Submission Package (SIP), an Archival Information Package (AIP) or a Dissemination Information Package (DIP) can be used.

---

[6] Meta-Informando is called this in Portugal, and in the Brazilian archival field it is called Metadados.

After this process there is the "digital challenge" according to Pinto (2010), which consists of two basic needs:

1. The need to guarantee intelligibility and continued access to information regardless of technological changes;

2. The inseparable need to identify the context of production this information and subsequent interventions.

Following the preservation perspective, elements were listed that involve the informational unit in the digital context, which is the multidimensionality that exists in them. These elements were chosen by Pinto (2009) and are subdivided as follows:

- Physical Dimension: This includes the way in which the document is stored (DVD, *CD, pen drive,* etc.), i.e. in this context, preservation is about having media that can be read and that are not obsolete. In this way, they are often not interpreted by *hardware* because they are already obsolete;

- Logical Dimension: The logical dimension is the organized part of the information, i.e. its structure and how it is organized, in GIF, TIF, DOC or PDF format. It is worth noting that it depends on a certain moment of being physically registered in order to be accessed, but it does not have a specific support;

- Conceptual Dimension: When we are in a digital environment, information is understood as code and it must have a meaning for the human being, so this does not occur in the logical or physical dimension, so digital signals are transformed into analog ones and can be understood in the final result in various formats - image, text, etc.

- Essential dimension: Set of essential elements that have been chosen for the preservation of information. Thus, the requirements presented above with regard to security levels are summarized in the metadata: descriptive, technical, administrative and structural, all of which are linked to each other according to the organic-functional structure.

- Therefore, it is worth noting that all this characterization is related to the information units in order to have integrated information without any alteration

Chapter 3

3 DIGITAL REPOSITORIES

In this section we'll look at the topic related to our aim of showing the historical journey to the *Open Archive Information System* (OAIS) conceptual model.

3.1 History

Researchers Vechiato and Fechal (2013) point out that the first repository appeared in 1990 in the United States, specifically at the *Los Alamos* National Nuclear Energy Laboratory, and was called ArXiv, which was used in the fields of computer science, physics, mathematics and non-linear sciences. The main objective of this repository was scientific communication, which at the time was to fill the gap created by the crisis in scientific journals.

> Digital repositories have emerged as an alternative for accessing, disseminating and preserving scientific production, which rose dramatically at the end of the 20th century. The *Open* Archives *Initiative* (OAI) has provided new possibilities for the process of scientific communication through the inclusion of open access institutional repositories with the aim of organizing, disseminating and providing access to scientific information (SHINTAKU; MEIRELES, 2010 *apud* VECHIATO; FECHAL, 2013, p. 105).

Barreto (2010) emphasizes that repositories in the context of information units first appeared in the university sphere and were joined by *Open Access* in relation to scientific literature, but even with this origin, which was born in the context of institutional production, they are also used for the following activities: archiving, disseminating and preserving other types of originals and documents, such as articles, among others.

Around 2002, the Massachutes Institute of Technology (MTI), together with *Hewlett-Packard,* launched an institutional format repository in which it had all the guidelines for open access and which expanded knowledge management in the scientific environment.

> In addition to expanding access to research, reasserting control over academic knowledge, reducing the monopoly of scientific journals, among other significant changes in the scientific communication system, it has the potential to serve as tangible indicators of a university's quality and to demonstrate the scientific, social and economic relevance of its research activities, increasing the visibility, *status* and public value of the institution (SANTOS et al., 2009, p. 223).

According to Santos et al. (2009), repositories can take different perspectives and forms with different subjects, communities and objectives, as their main focus is to preserve information within the system for users to access quickly and securely. "Integrating the problems and technical solutions relating to the preservation and authenticity of digital information, which makes sense to keep in view of its access to the public in different locations." (CORUJO, 2014, p. 54).

According to Pinto (2016), the fundamental characteristics of a repository must be authenticity, trustworthiness, integrity and intelligibility, i.e. these requirements are of the utmost importance for a reliable repository, as they advocate the following aspects:

- Authenticity: this parameter is used to check that the information is really safe and complies with legal requirements, and that an investigation is carried out in order to comply with this precept;
- Reliability: this aspect denotes whether the information is true, according to the context in which it was created;
- Integrity: aspect related to whether the information has not undergone any changes in its information flow, i.e. from its production to its arrival in the digital repository;
- Intelligibility: aims to maintain the ability to present the essential elements of authentic digital objects.

One of the main concerns regarding the preservation of information is security, which is inherent to the previous one and is the basic principle for creating a repository that meets all the preservation requirements.

Based on the study carried out by Pinto e Silva, in the field of Archivology, Organizers lack:

> [...] an approach that encompasses, from the conception phase of the technological platform (hardware and software), to the production, circulation, evaluation, storage, availability and preservation of information, the entire organization and its business processes. (PINTO; SILVA, 2014, p.[4]) .

The amount of use of repositories by organizations is more frequent and various types of repositories are emerging, of which information managers must be aware. As mentioned in the course of our research, technological evolution is a daily occurrence and repositories must be taken care of, since they are often treated as databases and security precepts are forgotten in such a way that there is the possibility of a huge loss of information in both analog and digital media. It's worth noting that digital repositories aggregate various types of information, and therefore specific *software* for them, as will be mentioned in the following sections.

> They encompass the scientific output of a given institution, most commonly research institutes and universities. They generally host a collection of research documents *(pre-prints* and *post-prints)*, although they can include technical reports, manuscripts, data, video clips and images, as well as administrative data to support the institution, such as the local documentation archive, theses, dissertations, books and others (BOSO, 2011, p. 35 *apud* ZUCATTO; RIBEIRO JÚNIOR, 2014, p. 6).

It is worth highlighting the functionality of digital repositories, since the main objective of all the areas that use this *software is to* use and access the information it holds.

3.2 Digital archive and digital repository

According to Barbedo (2005), in the 1970s concern arose about consolidating the preservation and accessibility of databases produced in the context of statistical studies, which gave rise to the first precepts about digital archives in which the first country to use this context was France, the Center for Contemporary Archives of France and the *Public Record Office* (PRO), implemented databases that had information in the context of digital archives.

Digital files have been worked on in different ways, in contexts in which each specific situation required it, and so their definition has become known as:

> The Digital Archive is therefore a structure comprising technology, human resources and a set of policies for incorporating, managing and making accessible digital objects of an archival nature on an ongoing basis. Archival information is distinguished from any other by the fact that it is produced for the primary purpose of providing evidence of an organizational activity (BARBEDO, 2005, p. 12).

In his doctoral thesis, Serrano (2013) points out that the digital archive can be considered a repository that brings together various subjects, including "gray literature", previous versions of documents or *pre-prints* and content related to scientific data. In this way, it encompasses various types of information relating to the digital environment.

Within the information units that use digital archives, we will now demonstrate how digital repositories are used in them and what their main uses are.

Following the perspective of Zucatto and Ribeiro Júnior (2014), information units in the digital sphere only exist due to Information and Communication Technologies (ICT) and their popularization, because over the years organizations have absorbed ICT and had to adapt to the new processes created, with this, repositories emerge in this informational context.

Continuing from this perspective, these repositories lead us to the idea of preserving information, because unlike paper media, digital media will be more secure with regard to this aspect of preservation, because they need a logic to be created, which is related to the architecture of information that provides a whole structured informational content for its users to be able to have proper access and satisfactory use within the system.

In the following sections, the respective concepts of our object of study and their types will be explored in greater depth. Lastly, in the work cited by Zucatto and Ribeiro Júnior (2014), we looked at the use of ICT in information units, up to and including repositories, as shown in the figure below.

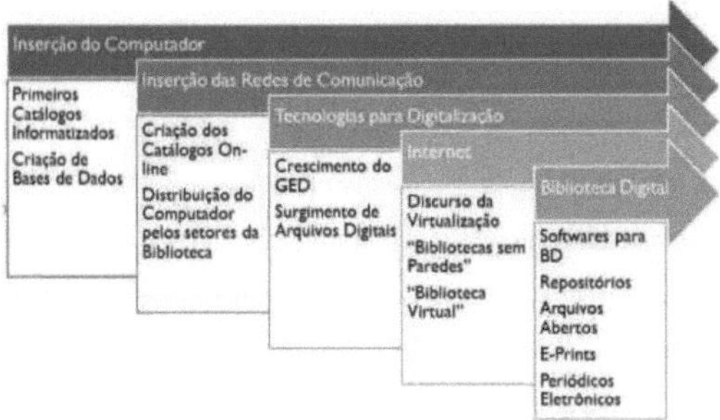

Figure 2 - Overview of ICT uptake
Source: Zucatto and Ribeiro Júnior (2014).

3.3 Conceptual model - OAIS

We will now outline the main thematic areas of our subject, which requires attention, as all the details are required to understand how repository *software* works and how it evolves.

To start specifying the conceptual model of the *Open Archive Information System* (OAIS), we need to talk about how this software came about. It all started in the 1990s, when a working group began to worry about the issue of digital preservation, sponsored by the RGP and the CPA, and its thinking was related to the issue of reliable digital archives. With the research sources presented in his master's thesis, Corujo (2014) emphasizes that these working groups were made up of the following organizations: NASA (United States), the *Centre National d'Études Spatiales in* France, the *British National Space Centre* (BNSC) in the United Kingdom and the European Space Agency (ESA), all of which have the necessary knowledge of all the previously mentioned levels of digital information.

The thinking of this group of organizations led companies in the field of information units to start thinking about the issue of digital preservation and after this research, analysis and *feedback* led to a more in-depth study of the issue until a solution was found.

The *Open Archive Information System* (OAIS) was created with the aim of standardizing the practice of digital preservation. According to Corujo (2014), the OAIS aims to offer a range of recommendations for the implementation of preservation programs, because it corresponds to a set of technical aspects of the life cycle of a digital object, which consists of: ingest, storage, data management, administration, access and preservation planning.

The model also addresses issues linked to metadata[*] , recommending five types of

---

[*] Metadata comes in the same sense as metadata, as it will be used in the following sections.

metadata for each digital object: referencing information (identification); provenance information (including preservation history), context, fixity (indicators of authenticity), and representation. The model also addresses issues related to metadata (CORUJO, 2014, p. 31).

Among the types and security requirements of digital repositories, the OAIS is considered a trust requirement, since it meets all the precepts and is used mainly by educational institutions, since it has the mission of offering long-term access and reliability to the digital resources that are being managed in relation to the location that has currently been designated. Therefore, we have to remember that the main function of the OAIS is to preserve the information that is stored, according to the technology of the time.

According to Corujo's research (2014), a reliable digital repository must have some attributes to follow this perspective:

> [...] compliance with the OAIS reference model, administrative responsibility, organizational viability, financial sustainability, technological and procedural adequacy, system security, procedural *accountability* [...] (CORUJO, 2014, p. 32).

When we refer to the OAIS as an open model, it is because it has been developed openly for all communities with the aim of making all the information within it available. Corujo (2014) argues that the OAIS in relation to archived information:

> The OAIS model provides a coherent view of the problem of archiving and digital preservation, defining a set of concepts and functions that are the basic elements essential for understanding it. To this end, it presents a complete terminology that is not intended to replace existing terminologies, but aims to keep the model at a general level of abstraction regardless of any specific application context, to make it reusable and applicable to all areas, sectors and types of digital information. (CORUJO, 2014, p. 65).

The OAIS issue also involves information packages, which are called as follows: *Submission Information Package* (SIP)*; Archive Information Package* (AIP) and *Dissemination Information Package* (DIP).

Chapter 4

4 TYPES OF REPOSITORIES

Having presented the main functionalities and general characteristics of Digital Repositories, we will now move on to the existing typologies within this segment, starting with Institutional Repositories. These can basically be conceptualized as digital collections that aim to store, preserve, disseminate and allow access to the scientific and intellectual production of university communities (RODRIGUES et al., 2004; VEIGA et al., 2016). As such, these repositories are better suited to carrying out these tasks with data that is limited in scope to the academic community.

By putting these principles into practice, repositories help to increase the visibility and added value of the institution, adding quality and demonstrating the scientific, economic and social relevance of the academy.

> An institutional repository includes a variety of materials produced by professors\researchers, from a variety of documents, such as *e-prints,* technical reports, theses and dissertations, data sets, and teaching materials. Some institutional repositories are also being used as electronic publishing houses for the publication of e-books and electronic journals (ROSA, 2011, p. 136).

According to Rodrigues et al. (2016, p. 73) institutional repositories are considered by some authors to be one of the most important elements of archives and digital libraries, "they generally have well-defined common concrete objectives identified with the organizational strategy. They seek to aggregate all the scientific outputs of their institutions and do so in open access [...]".

Before institutional repositories were conceived, there were initially repositories characterized as thematic, which were described as specialized digital repositories with works relating to a particular branch or subject. However, this characterization has since evolved and repositories have begun to be grouped together and placed under the responsibility of an institution. However, we can still say that institutional repositories are made up of thematic micro-repositories.

Currently, institutional repositories have managed to share the market for scientific publications with specialized journals after the consolidation of initiatives such as *Open Access,* which has been building an environment conducive "to free access to scientific production in a legitimate way, changing not only the process of acquiring scientific information, but also its production, dissemination and use" (WEITZEL, 2006, p.52).

Another existing type of repository that we are going to look at is *E-learning,* which consists of information in the educational context of Tec Minho and the University of Minho, the aim of which, according to Carvalho et al:

- To contribute to the preservation and dissemination of the didactic and pedagogical production of the University of Minho and TecMinho;

- Valuing didactic and pedagogical development;

- Preserving the University's intellectual memory;

- Give free access to the educational content produced;

- Storing Educational Content;

- Encouraging the adaptation and reuse of content;

- Develop the community's spirit of sharing.

In view of this, a repository that has all these functions aims to group all these functions together in one place and relate them to the learning objects that support the learning management platforms used by the two higher education institutes presented above in the concept of *e-Learning*.

In the document called Educational Content Repositories by Carvalho et al. (2012), they present that it uses the DSPACE platform which was practically built for the storage, maintenance and collection of digital materials, in which those responsible for this platform were the *Massachusetts Institute of Technology* (MIT) and *Hewlett-Packard Labs* (HP), both through their library services were responsible for the development of DSPACE.

Still on the subject of *E-learning, it should be* noted that it includes documents from various contexts in digital format, which are the result of pedagogical activities by the institutes and are the following types according to Carvalho et al. (2012): presentations, exercises, questionnaires, Packages, teaching sheets, summaries, etc. The hyperlink to access the repository is: https://e-repository.tecminho.uminho.pt, and there is also a document[9] which presents all the content related to this typology.

---

[9]

General description of the Digital Repository, which presents all the levels involved in *E-learmng,* which is available at the following address at:<https://elearmng.iefp.pt/pluginfile.php/49476/mod_resource/content/0/produtos/Descricao_Geral_do_Repos itorio.pdfX

Figure 3 - e-Leaming Repository
Source: Carvalho et al. (2012).

In this context, data repositories have emerged, which according to Peixoto (2012), are interconnected with the *data sharing* movement, which is nothing more than the possibility of this data being shared with other researchers, in other words, an interoperability that exists within this system, since it complies with the requirements of the information flow and now has a different level, which is communicating.

According to Peixoto (2012 *apud* TORRES-SALINA, 2012), there are two ways of sharing data in this environment: formal and informal. The first is the one that plays a fundamental role in recovering information and the other directs us in a different way, which is information shared through bureaucratic means (workshops, meetings, research groups and researchers' personal websites), which makes it more difficult to publish this material in public.

The use of this repository is well accepted in the United States of America and Europe and is, in a way, being used in a consolidated way with other repositories of the same style.

According to the data available on Portugal's Open Access Scientific Repository (RCAAP), it has this type:

> [...] marks the beginning of its intervention in the field of curation of data resulting from research, its organization in scientific data repositories and its access. A pilot project is underway in this area with the aim of developing some practical activity in the area of curating and sharing scientific data (RCAAP, 2017, *Online).*

With this, we denoted the evolution of this type and saw what concerns it consists of within the organizational context.

4.1 Existing platforms

According to Sayao et al. (2009), it is only possible to develop institutional repositories that meet the technical and functional requirements, specified models and that are also adaptable to the diversity of interests existing in the user profile, if they adhere to software platforms that allow expansion and integration with other software, making it possible to meet current and future demands.

Due to the increase in the volume and diversity of materials in digital format, it is essential that the technological platform is effective in supporting the performance of digital repositories.

Currently, there are excellent software options for implementing digital repositories, with the most excellent, both technically and in terms of functionality, being free and open source software. However, the choice of platform to be used depends on the characteristics of the repositories, since the software platforms available were developed in different contexts and have specific objectives and vocations.

The form of storage is not the only concern in terms of digital preservation. The ability to ensure that information remains accessible even if the technological platform on which it was created becomes obsolete is just as important as the first, because "digital preservation is the activity responsible for ensuring that communication is possible not only across space, but also across time" (FERREIRA, 2006 *apud* CASTRO et al., 2009, p. 284).

Most *open source* platforms work through the OAI-PMH protocol, which increases the visibility of the contents of the collection and allows communication, interoperability and integration between different repositories and information systems.

4.2 Digital Archival Repositories

As part of the process of describing our object of study, we have come to the part of showing digital repositories in the archival environment, which have a direct relationship with the principles of archivology. It's good to highlight this relationship in order to emphasize the specific part of the digital archival repositories that will be presented here: Repositorio de Objetos Digitais Auténticos (RODA) and Archivematica.

When we look at the history of digital repositories, we can already see some relationship with archivology in terms of the principles for having a reliable archival repository (authenticity, trustworthiness, integrity and intelligibility) and, corroborating this fact, the authors Vechiato and Fernal (2013) address the perspective that digital repositories are connected to the information environments of archives based on the association of archival principles with repositories, a relationship discussed in the table below:

Table 03 - Relationships between archival principles and repositories

Archival principles	Digital Repositories
Provenance	They promote the convergence of digital archives. This centralization will safeguard the archives produced by a particular institution or people so that they are not mixed with the archives of other funds.
Organization	They present a hierarchical structure that represents administrative structures and organizational functions.
Uniqueness	Each archive is deposited in a specific location, which is part of the document structure.
Archival integrity	Promotes the preservation of digital files.
Cumulativeness	Progressive, natural and organic formation.

Source: Vechiato and Fernal (2013).

Following this tenuous line, we can see how digital repositories are parameterized since "they have the fundamental elements for the development of archival practices, configuring themselves as a possibility for the archivist, who needs to be aligned with the development of ICT" (VECHIATO; FERNAL, 2013, p. 118).

4.2.1 Repository of Authentic Digital Objects (RODA)

With this introductory overview of the relationship between archives and repositories, we will now look at two pieces of software that are used in this connection: RODA and

Archivematica.

The Repositorio de Objetos Digitais Auténticos (RODA) is an open-format *software program* whose purpose, according to Silva (2014), is to create the probability of preserving information in digital format. This happens with a program that makes it possible to aggregate the main management and preservation references for archival funds.

> Roda is a project that aims to develop and promote technological solutions, finalizing in the construction of a prototype digital repository capable of incorporating, describing and giving access to all types of digital information produced in the context of Public Administration (FERREIRA et al., 2007, p. 1).

Ferreira et al, (2007), report that the digital archive will be planned within RODA from the perspective of achieving a system capable of making all the functions of a digital archive certain, in accordance with the OAIS that we refer to in the course of the research, ranging from the management to the dissemination of the information produced within the system.

It is worth noting that RODA is made up of a package of programs and services: *Submission Information Package* (SIP)*; Archive Information Package* (AIP) and *Dissemination Information Package (*DIP) which, in turn, is capable of building a reliable archival repository. In figure 04 we have the repository's website:

Figure 4 - RODA's *World Wide Web (*WEB) page interface

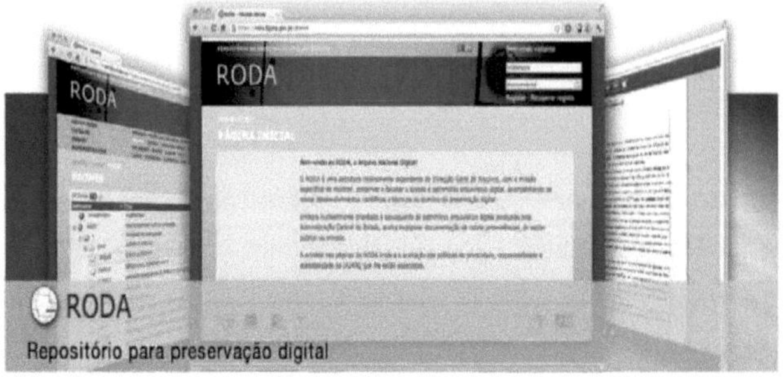

Source: RODA(2017).

According to Barbedo et al. (2009), RODA's main objective is to ensure the authenticity of digital documents, which is why it has been standardized based on international specifications, such as the *General International Standard Archival Description (ISAD (G))*, in order to become a reliable repository due to the high standard of security and management of the information it holds.

In the context of RODA, following Silva's (2014) perspective, we have the inclusion of managing and giving access to various materials in digital format, which are mostly produced by large public organizations, and this software is made up of *opensource* technologies and is supported by the OAIS, EAD, METS and PREMIS standards.

"RODA's administration mode allows repository managers to edit descriptive metadata, launch preservation actions...] control access by users, consult statistics, access logs, among many other options" (RODA, 2017, online).

Finally, we list the organizations that can use RODA in our country, according to the National Council of Archives of Brazil (2014), which establishes the main requirements for implementing repositories in institutions that are part of the National Archives System of Brazil

(SINAR):

- National Archives;
- Archives of the Federal Executive Branch;
- Archives of the Federal Judiciary;
- State Archives of the Executive, Legislative and Judicial Branches;
- Archives of the Federal District Executive, Legislative and Judicial Branches;
- Municipal Archives of the Executive and Legislative branches;
- Individuals and companies can join SINAR when they hold files related to the central body.

With all that has been mentioned, we have seen the importance of RODA for the preservation of digital objects and all the relationship it has with Archivology, and we highlight the issue of characterization from its ingestion into the system to the dissemination of information, which follows archival parameters.

4.2.2 ARCHIVEMATICA

Now we're going to introduce another piece of *software* in the area of digital archival repositories, Archivematica, which is free *software* used to create digital repositories. The researcher Santos et al (2014) reports that it was built in an open source format in which you have access to the source code[10] , created by the company *Artefactual System.*

According to the information investigated on the repository's *website*, we detected the project's partners, including, according to Santos et al. (2014), the United Nations Educational Scientific and Cultural Organization (UNESCO), the Vancouver Archives in Canada and the *British Colombia* University:

> [...] Archivematica in particular should consider long-term preservation strategies, avoiding technological obsolescence, incompatibility of formats, versions and media. Ensuring that digital documents remain authentic, accessible and usable over the years (SANTOS et al., 2014, p. 72).

Continuing in this vein, we'll look at the steps involved in preserving and accessing a repository:

Table 04 - Activities carried out by the information package for preservation and access

Stages	Activity

---

[10] Available at:<https://www.archivematica.org/en/>

Submission Information Package (SIP)	Information package submission, the first stage after installing the repository, where the producer will submit the information packages, transferring the files to the repository. At this stage Archivematica will highlight in the *"transfer"* menu bar that the transfer is taking place. After the transfer, the packages will be submitted to *"ingest"*, the stage at which the packages are inserted, i.e. submitted to the repository. This is when the microservices are performed, converting formats before being archived.
Archive Information Package (AIP)	Archiving Information Package, in this second step, the administrator will give the command to archive the information package previously submitted. At this stage Archivematica will highlight *"ingest" in* the menu bar, indicating that the package is being archived[...]. It should be noted that the repository accepts multiple file formats, including text documents, music, videos, images, plans and so on;
Dissemination Information package (DIP)	In this third and final stage, the ostensible information is disseminated using *the ICA-AtoM software,* which stands for Access to Memory, software for describing and disseminating archival documents of a permanent nature, developed by the same company *Artefactual System*. It is content management software based on standards for archival description and the International Archives Standards of the International Council on Archives (CIA). It's worth noting that the CIA itself commissioned *Artefactual System* to develop specific *software* for archival description based on its international standards and also to promote the dissemination of and access to the information described.

Source: Adapted from Santos et al. (2014).

With the packages and their activities presented, it is worth highlighting how they work, and for this, the Brazilian Institute of Information on Science and Technology (IBICT) provides information in two manuals called Guia de instalando e configurando Archivematica / AtoM by Novais et al. (2017), which aims to:

This guide is aimed at Information Technology (IT) professionals with intermediate knowledge of Linux operating systems, so that it can be interpreted successfully. It is the result of a collective effort shared by all the authors identified in this work, aimed at guiding public and private institutions in Brazil to install and configure the free software Archivematica, version 1.6, and AtoM, as its access interface. The steps for installing and configuring the software are clearly presented, from preparing the environment to the step-by-step process for installing and configuring both (NOVAIS et al., 2017, p. 7).

Santos et al (2014) permeate another fact highlighted by the standards used within Archivematica, the International Standard for Archival Description (ISAD (G)) used to standardize the archival description of software, the International Standard for the Registration of Archival Authority for Collective Entities, Persons and Families (ISAAR (CPF)) is used to guide the preparation and standardization of archival authority records to provide guidelines for the preparation and standardization of archival authority records, as well as the International Standard for Describing Institutions (ISDIAH), the International Standard for Describing Functions (ISDF) and others that are emphasized in the research.

4.3 Library repositories

Continuing the presentation of the types of digital repositories, we will now list the library repositories that are characterized in the institutional, more specific part that consists of what software exists in the area.

Before we start dissecting the issue of library repositories, we have to differentiate between two words that Pinto (2016) makes very clear in his material, which are the differences between *open access (*access to information) and *open sources* (platforms/software), as presented in the previous sections, aim to carry out the entire information flow to disseminate the digital intellectual production of university communities, while the second is directly linked to the logical part which is the construction of software to carry out and manage the functions of digital repositories, in the scope of the research we will address: Dspace and *Fedora.*

4.3.1 DSPACE

According to Pinto (2016), DSPACE was created by MIT thanks to the partnership between the MIT libraries and *Hewlett Packard Research labs in* which it is aimed at information in digital media in relation to research, and it was used as an *open source* solution, as it has the levels of access, management and production of materials in the academic context.

Remembering that DSPACE has several functions to be carried out and requirements in which it has also been made available since 2002. According to the material presented by Pinto (2016) at the Information Science Days, she parameterized the following DSPACE activities: capture, description, distribution and preservation, which aggregate the following activities:

- Capture: sees what information is digital and who its producers are, i.e. collects the initial data produced within the system.
- Description: describes the information at technical, administrative and other levels. In which the description of the document is considered in analog form, but now in digital form, and with the help of automatic tools;
- Distribution: characterizes where this information can be made available and by what means, whether within the company, on the Web or other dissemination mechanisms;
- Preserving: Stores information over a long term according to a stable scale.

Figure 5 - A dynamic, *open source* digital repository

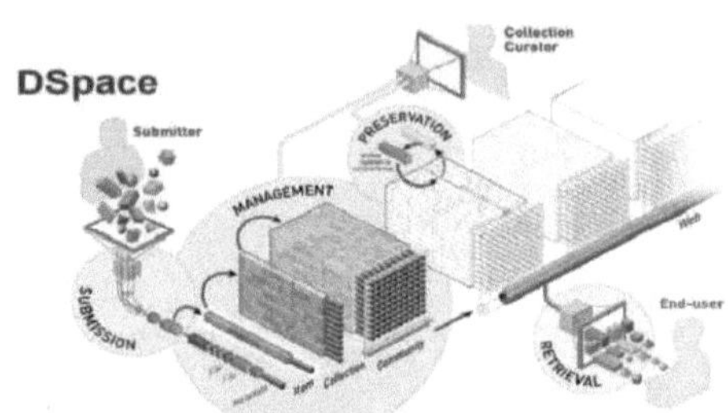

Source: Pinto (2016).

The DSPACE information model is divided into communities that list the production units, the academic or research community and colleges that direct the type of existing items, items that consist of an information unit.

4.3.2 FEDORA

Following the presentation of library repositories, we come to *FEDORA, which* is a repository with an open source format and aims to manage and disseminate content in digital media. According to the FEDORA website (2017):

> It is especially suitable for digital libraries and archives, both for access and preservation. It is also used to provide specialized access to very large and complex digital collections of historical and cultural materials, as well as scientific data. Fedora has an installed user base worldwide that includes academic and cultural heritage organizations, universities, research institutions, university libraries, national libraries and government agencies (FEDORA, 2017, online).

According to Carvalho and Oliveira (2011), this type of repository is made up of content that integrates data and metadata and in relation to the behavior or associated services that can be used for the digital object[11] . *Fedora* consists of four parts presented in the context of this research

- Digital object identifier: provides a persistent unique identifier for the digital object;
- Descriptive Perspective: mentions the main metadata needed for management, retrieval and relationship with other objects, within *Fedora* the *Fedora Object XML (FOXML)* metadata is used.
- Item perspective: it is intended to represent the main content of the digital object (texts, images, audio, video, etc.).
- Service Perspective: this is the mechanism for associating behavior or services that makes it possible to define the representations of the exteriority of *Fedora*'s digital objects.

Figure 6 - Fedora's digital object architecture

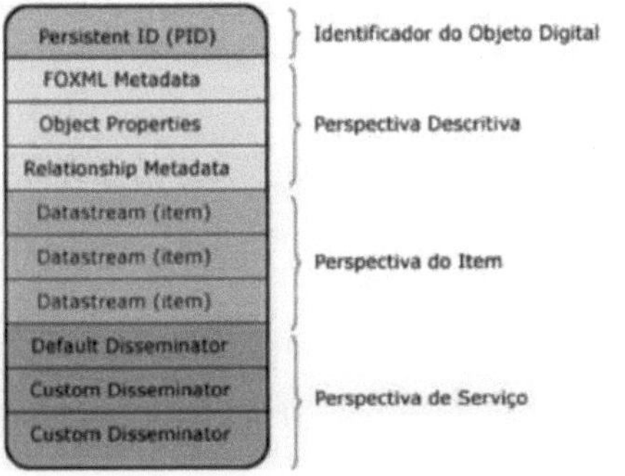

Source: Carvalho and Oliveira (2011).
Source: Carvalho and Oliveira (2011).

It is managed by the *Fedora Leadership Group* and has several activities that qualify it as qualitative aspects and that propagate its service. This company has evolved over time, from 1997 to the present day, all of which is shown on its website[12] .

---

[11] According to Thibodeau (2002) it is an information object of any type and format in the digital environment. He proposes that they are subdivided into: physical, logical and conceptual.
[12] http://www.fedora-commons.org/about

Chapter 5

5 THE SECURITY PARALLELS WITHIN SOFTWARE

In view of all the information processing that we have covered in this research, we will now begin a section that depends on the entire information flow, because from the moment the information enters the system, whether it is analog or digital, there must be concern about the quality of this information and this adds to the issue of information security. In addition to the requirements mentioned above of authenticity, reliability, integrity and intelligibility, we will now look at two other aspects of information security: its certification and how this preservation can be kept secure.

There are various tools that enable us to have secure information. They are all due to the evolution of Information and Communication Technologies (ICTs), because the more complex ICTs are, the better information security should be.

According to Santos (2015), all repositories have requirements for their proper reliability in order to avoid threats and damage to their system: monitoring, planning and constant maintenance so that the system does not become obsolete according to the evolution of due technology and all this leads to an organizational awareness, in which it enables the information manager to have work in this area constantly providing a motivation conscious actions and an implementation of the strategy adopted to achieve its mission in the field of digital preservation.

"All those who deposit information in the repository will have to rely on a collaborative digital preservation environment so that this information is preserved and its long-term accessibility guaranteed" (SANTOS, 2015, p. 83).

After all the basic contextualization on repository security, we come to the culmination of the session in which we will present the issue of certification and preservation of information. Following Thomaz's (2007) tenuous line of thought, certification within digital repositories has become essential, as it is used as a mechanism and a measuring instrument, which makes it possible to retrieve information quickly and to disseminate and implement good practices within the system.

According to Thomaz (2007), information certification has three pillars, which are the following requirements: organization, technology and management, as shown in figure 5:

Figure 7 - Three Pillars of Digital Certification

Source: Thomaz (2007).

The pillar involving the organization consists of: "certifying the charge, scope, objectives, financial availability and commitments of an organization to engage in digital preservation" (THOMAZ, 2007, p. 86). In other words, this pillar is related to bureaucratic decisions within repositories, leading to commitments involving adherence to standards related to digital management and preservation, such as the ISO 15.849 standards on *Information and Documentation - Record Management*.

Thomaz (2007) emphasizes in his article that other aspects such as those listed below meet the organizational requirements:

- Studies on the costs of digital preservation, such as the economic models proposed by *Brian Lavoie8*;
- Research on cost models and structures by *Shelby Sanett9*;
- An approach developed by the National Library of the Netherlands that establishes a tool for comparing the costs of migrating and emulating over time;
- More comprehensive cost formula proposed by the *Life Cycle Information for ELiterature* project (LIFE11).

It's worth noting that these aspects mentioned above give an idea of how high the cost of preserving information is, each one is related to the context in which it is inserted, but they are not the same in a place that the community itself builds on the references of organizational issues.

Another pillar that stands out is the technological one:

> The technological pillar certifies the adequacy of the repository's technical infrastructure and its ability to meet the demands of managing and securing the repository and digital objects. It involves hardware, software, storage media, networks, security measures, workflow managers, protocols, documentation and both technical and administrative skills (THOMAZ, 2007, p. 86).

As well as the technological requirements, there are also other parameters[13] that are

---

[13] Link to access Thomaz's technological requirements:

addressed in the technology and are presented in the research, and now we have to remember that organizations must have a diagnosis of what needs to be done to be able to see what the best solution is, with the examples of the following places: such as the National Archives of the United Kingdom15, or hiring a digital preservation service provider, such as *Berkeley Electronic Press* (bepress16) or the OCLC Digital Archive17.

Lastly, we have the certification parameter, which we will now address in terms of management, which "certifies the functions, processes and procedures needed to manage the digital objects themselves, grouped according to the functional entities defined in the famous SAAI reference model" (THOMAZ, 2007, p. 87).

With all these requirements for certification, it's important to check that there are models to follow, and to take them into account: The *Trusted Digital Repositories, Archival Workshop on Ingest and Identification, and Certification Standards (*AWIICS*).*

Having finalized the issue of certification, we now move on to another level, which is the preservation that is essential at this stage. According to Pinto (2009): "one can place in the year 2003 the main alert to the reality of information in digital media and the imminent risk of its loss and, consequently, of the loss of collective memory at scale [...]", with this concern for the preservation of information arises at the 32° Session of the UNESCO Conference the Charter for the Digital Preservation of Digital Heritage, which aims to:

> [...] in unique resources resulting from the knowledge or expression of human beings, comprising cultural, educational, scientific or administrative resources and technical, legal, medical and other types of information, which is generated directly in digital format or from the conversion of analog material that already exists (PINTO, 2008, p. 5).

This concern begins with the physical support, where care must be taken to ensure that it does not become obsolete and the information is lost. Another point that comes to the fore is the question of the existence of a digital preservation plan in which each professional must have their responsibility, and then this procedure will actually take place.

Pinto (2008, p.128) covers the distribution by levels from "copying, modifying, preserving and making available resources, to guaranteeing the authenticity, protection and continued access to information [...]", with this we have seen that there is a range of requirements for good preservation, which depends on the division of tasks to be efficient, these ideas lead to having information accessible in a continuous way by formalizing it in a structure that is composed of one of the assignments or procedures that are within the information life cycle.

Remembering that we have to have continuous long-term preservation using all possible

---

strategies, remembering that they are framed in individual contexts because they depend on the situation in which the organizational unit is located, so libraries, archives, museums and document centers have totally different characteristics, but they need precepts to have good preservation.

Still in this context, Pinto (2008), presents the following topics on the preservation of this information within each information unit, but we highlight the archive in which all types of documents are received, have probative value, and fulfill all the requirements in which they have a specific preservation in which to scale the four levels of digital objects, as mentioned in previous sections;

> This last aspect shows that in addition to the issues raised in terms of building up the collections/sealing the information to be acquired/collected by the various institutions, there is also the issue of representing the information, or if you like, the processes and the environment in which the creation of meta-information takes place and the models/schemes used, as well as the question of authenticity, not ignoring the substantial differences in the positioning of professionals, the diversity of standards, the parallel paths that are taken and the impact that these behaviors have on crucial aspects of digital preservation (PINTO, 2008. p. 10).

Other forms of preservation that we should emphasize are the issue of regular encryption and the levels of access to information that is managed by the Institution's Information Technology System, according to the company's needs. Finally, a solution that must be taken with each technological evolution is to back up and adapt the software to each format, i.e. constantly update it so that it does not become obsolete and the information is kept for the long term, regardless of its context.

Therefore, we mention the forms of preserving and certifying the information, in which they require a common precept from the information manager in order to be done in the best way, it is worth mentioning that the IM in this aspect works integrated with other areas of knowledge and having an interdisciplinarity of various contexts.

FINAL CONSIDERATIONS

After presenting the history of digital repositories and their characteristics, we can see how they are made up of a wide range of knowledge, as they require a lot of specifics to be able to understand their entire theoretical context and their applicability, leading us to list the main issues within this research.

Its emergence has facilitated information processing, as we have seen in the course of this work, with a view to accessing and preserving the information it holds. However, we have seen that the structures added to them have begun to become more complex as technology has evolved, and this has led to the emergence of specific repositories for each information unit, in the context of this work we have listed the repositories used in archives and library science.

The repositories investigated in the field of Archivology, led us to think about the relationship between the archival principles and them, with which we found the literature pertinent to this context, and further corroborated our object of study. The *software used* in this investigation was RODA and Arquivemática, both of which have different functionalities. The former creates the probability of preserving information in digital format, while the latter must take into account long-term preservation strategies, avoiding technological obsolescence and incompatibility of formats, versions and media.

In the field of library science, two pieces of software were listed: DSPACE, which is linked to access to information in digital format and is widely used by libraries to access scientific information, and FEDORA, which is still little used in libraries, but has a range of functions from access to preservation, and can also be used in digital archives.

Having described the applicability of this software in the two areas of knowledge, we now turn our attention to the relationship between archivists and the types of repositories that we characterized in the research: RODA, which is classified as an institutional repository and is interconnected with the FEDORA platform, because its activities are directly linked to the preservation of archival collections; Arquivemática is also characterized as institutional and is associated with DSPACE.

The challenge of this research was to meet all the objectives, because as it is a bibliographical study, we searched for authors in the international and national context and this enabled us to broaden the view of our object of study and to describe it with quality literature and to expose synthesized research in the area.

With our objectives achieved, we can see that the digital repository is very important in information processing. With this, we bring it into the archival perspective, in which we identify a direct relationship, since it is a tool that has been built from a whole systematization in which the user has quality information.

Our suggestion for future research on the subject is to expand on the description of how software is used in this field of knowledge, from its creation to its use with users.

Therefore, digital repositories have emerged with the main aim of making information open to the entire community, since the beginning of scientific communication in the Second World War and we realize within the research that we have been able to detect this entire path, it is worth mentioning that this study is only the beginning of a research on the subject, which due to technological developments is expanding and causing new researchers to emerge on the subject.

REFERENCES

ARAÚJO, C. A. A. de. Fundamentals of Information Science: theoretical currents and the concept of information. **Perspectivas em Gestao & Conhecimento**, Jodo Pessoa, v.4, n.1, p.57-79, jan./jun., 2014.

BARRETO'S blog. [S.l.]. 2010. Available at: <https://aldobarreto.wordpress.com/2010/04/21/bases-de-dados-e-repositorios-de-informacao/>. Accessed on: 07 Oct. 2017.

BARBEDO, F; CORUJO, L; CASTRO, R; FARIA, L; RAMALHO, J. C; FERREIRA, M. **Roda:** repository of authentic digital objects. 2007. Available at: Accessed on: 27 Oct. 2017.

BATISTA, AnaAlice et al.. **General description of the e-Learning Repository**. [20--]. 29 p. Available at: <https://elearning.iefp.pt/pluginfile.php/49476/mod_resource/content/0/produtos/Descricao_Geral_do_Repositorio.pdf>. Accessed on: 08 Oct. 2017.

CASTRO, C. Y. H. de et al. Trusted institutional repositories: Institutional repositories as a tool for digital preservation. In: SAYAO, Luis et al. (Org.). **Implementing and managing institutional repositories**: policies, memories, open access and preservation. Salvador: EDUFBA, 2009. p. 283-304. Available at:<https://repositorio.ufba.br/ri/bitstream/ufba/473/3/implantacao_repositorio_web.pdf>. Accessed on: October 10, 2017.

CARVALHO, C.L.OLIVEIRA, R.R **Digital Libraries and the Fedora Repository.** Federal University of Goiás, 2011. Available: at:<http://www.inf.ufg.br/sites/default/files/uploads/relatorios-tecnicos/RT-INF_002- 11.pdf>. Accessed on: October 10, 2017.

CONCEIQAO, A.S. **Informando Arquivística:** O insumo da sociedade contemporánea- a riqueza das organizanóes. Archeion, Joao Pessoa, July/December 2013. Available at: <http://periodicos.ufpb.br/ojs/index.php/archeion/article/view/17140/9755>. Accessed on: October 17, 2017.

CORDIS. **Digital Repository Infrastructure Vision for European Research**. 2016. Available at: <http://cordis.europa.eu/project/rcn/86426_en.html>. Accessed on: October 10, 2017.

CORUJO, Luis Miguel Nunes. **Digital Repositories and Confianna** - An example of a Digital Preservation repository: RODA. 2014. 255 f. Dissertation (Master of Science in Documenting and Informing) - Faculty of Letters, University of Lisbon, Lisbon, 2014. Available at: <http://repositorio.ul.pt/bitstream/10451/18109/1/ulfl179121_tm.pdf>. Accessed on: October 10, 2017.

DeltCI - Electronic Dictionary of Information Science Terminology. Available at: <https://paginas.fe.up.pt/~lci/index.php/1648-investigar/deltci-dicionario-eletronico- terminologia-ci>. Accessed on: October 28, 2017.

DURASPACE. **Who we are**. Available at: <http://www.fedora-commons.org/>. Accessed on: November 29, 2017.

EPRINTS. **About us**. 2017. Available at: <http://www.eprints.org/uk/index.php/about/>. Accessed on: Nov. 12, 2017.

EUROPEAN. 2017. Available at: <http://pro.europeana.eu/about-us/>. Accessed on: 10 Nov. 2017.

FEDORA. **Who we are**. Available at: <http://www.fedora-commons.org/>. Accessed on: November 29, 2017.

FLORES, D. **Preserving digital archival information:** repercussions for cultural heritage. Em Questao, Porto Alegre, Jan/Jun. 2013. Available at:<http://seer.ufrgs.br/index.php/EmQuestao/article/view/27024/31549>. Accessed on: October 15, 2017.

FLORES, D; SFREEDO, J.M.A. **Archival information security:** access control in state public archives. Perspectivas em Ciencia da Informando, v.17, n.2, p.158-178, apr./jun. 2012. Available at: <http://www.scielo.br/pdf/pci/v17n2/a11v17n2.pdf>. Accessed on: October 5, 2017.

GAUTHIER, F. C; YAMOKA, E.J. **OBJETOS DIGITAIS: EN BUSCA DE LA PRECISIÓN CONCEPTUAL:** IN **SEARCH OF** CONCEPTUAL **PRECISION.** Londrina, 2013. Available at: <http://www.brapci.ufpr.br/brapci/_repositorio/2014/12/pdf_3b9115db0a_0024408.pdf>. Accessed on: October 28, 2017.

GIL, Antonio Carlos. **How to prepare research projects.** 4. ed. Sao Paulo: Atlas, 2002.

GUERRA, E.L.A. **Manual of Qualitative Research.** Anima Educando, 2014. Available at:< http://disciplinas.nucleoead.com.br/pdf/anima_tcc/gerais/manuais/manual_quali.pdf>. Accessed on: October 22, 2017.

GUIMARAES, A.T.R. ALMEIDA, F.A.S. SILVA, A.M. **O modelo quadripolar aplicado a educando mediada por tecnologia da informando e comunicando:** um estudo empírico. Prisma, 2011. Available at:http://revistas.ua.pt/index.php/prismacom/article/viewFile/1247/pdf. Accessed on: October 15, 2017.

IBICT. **About digital repositories.** Available at: < http://www.ibict.br/informacao- para-ciencia-tecnologia-e-inovacao%20/repositorios-digitais/ sobre-repositorios-digitais>. Accessed on: October 3, 2017.

NOVAIS, MARCOS et al. **AtoM INSTALLATION AND CONFIGURATION GUIDE**

Archivematica. Brasilia, 2017.Disponivel em: <http://livroaberto.ibict.br/bitstream/123456789/1067/2/Archivematica%20-%20guia%20de%20instala%C3%A7%C3%A3o%20e%20configura%C3%A7%C3%A3o.pdf >. Accessed on: October 28, 2010.

OPEN ACESS. **Driver II** - Digital Repository Infrastructure Vision for European Research. University of Minho. 2017. Available at: <http://openaccess.sdum.uminho.pt/?page_id=222>. Accessed on: September 10, 2017.

PAES, M.L. **Arquivo:** teoria e prática. Rio de Janeiro: FGV, 2007.

PEIXOTO, N.M. **Information architecture in scientific data repositories: analysis of the PELD** - Programa de Pesquisas Ecológicas de Longa Duracao **repository interface.** 2012. 78 f. Work (Course Conclusion) - Faculty of Library Science and Communication, Federal University of Rio Grande do Sul, Porto Alegre, 2012. Available at: <http://www.lume.ufrgs.br/handle/10183/69727>. Accessed on: September 8, 2017.

PINTO, M._.**Informa$ao em meio digital**: genr para preservar. Observatorio de Ciencia da Informaqao da Universidade do Porto, 2009. Available at: <https://paginas.fe.up.pt/~lci/index.php/251-investigar/eventos-nacionais-ci- uporto/jornadas-ciencia- informacao/515-jornadas- ciencia- informacao?showall=&start=7>. Accessed on: 29mar. 2017.

. **Preservation of Information** *[PowerPoint slides]*. 2016. Available at:<https://sigarra.up.pt/flup/pt/conteudos_geral.ver?pct_pag_id=1009057&pct_parametros= pv_ocorrencia_id=382667&pct_grupo=48554#48554>. Accessed on: March 29, 2017.
______________ . **Preservation in digital media**[*Power Point slides*].2016. Available at: < https://sigarra.up.pt/flup/pt/conteudos_geral.ver?pct_pag_id=1009057&pct_parametros=pv_o correncia_id=382667&pct_grupo=48554#48554 >. Accessed on: March 29, 2017.
. From document preservation to information preservation. In: **Document preservation and restoration in the post-custodial era.** 2008. p.127-196. Available at: <https://sigarra.up.pt/flup/pt/pub_geral.pub_view?pi_pub_base_id=77027>. Accessed on: March 29, 2017.

PINTO, M.M; SILVA, A.M. Um Modelo Sistémico E Integral De Gestao Da Informando Nas Organizanóes. **CONTECSI USP - International Conference on Information Systems and Technology Management.** Sao Paulo, Oct. 2014. Available at: <http://ler.letras.up.pt/uploads/ficheiros/3085.pdf>. Accessed on: 01 Apr. 2017.

RAMALHO, F; VEIGA, A. L. **EAD and EAC Meta-Information**. Faculty of Engineering of the University of Porto, 2013. Available at:<https://paginas.fe.up.pt/~ci05020/trabalhos/LCI_PreCon_AnaLuisaVeiga_FilipaRamalh o_relatorio.pdf>. Accessed on: March 29, 2017.

RIBEIRO JUNIOR, D. I.; ZUCATTO, A.C. P. Libraries and digital repositories: reflections, technologies and applications In: **XVIII Seminario Nacional de Bibliotecas Universitarias,** 2014. Available at: <https://www.bu.ufmg.br/snbu2014/trabalhos/index.php/sn_20_bu_14/sn_20_bu_14/paper/vi ew/638/270> Accessed on: 04 Apr. 2017.

RIBEIRO, M.F; PINTO, M.M. **O acesso aberto a investigado em ciencia da informado em Portugal**: alcance e impacto. 2009. Available at: <https://repositorio-aberto.up.pt/bitstream/10216/57297/2/73329.pdf>. Accessed on: 06 Apr. 2017.

RODA. **WHO WE ARE.** 2017. Available at: < http://www.uel.br/revistas/uel/index.php/infoprof/article/view/17272/pdf_1.> Accessed: Oct. 28, 2017

RODRIGUES, Eloy et al. **RepositóriUM:** creation and development of the Institutional Repository of the University of Minho. In: CONGRESSO NACIONAL DE BIBLIOTECARIOS, ARQUIVISTAS E DOCUMENTALISTAS, 8., 2004, Estoril. **Proceedings...** Estoril: Portuguese Association of Librarians, Archivists and Documentalists, 2004. Available at: <https://repositorium.sdum.uminho.pt/bitstream/1822/422/1/BAD_artigo%20-%20Final.pdf>. Accessed on: March 28, 2017.

RODRIGUES, M.E et al. The repositories of Portuguese higher education institutions: a comparative study. In: CONFERENCIA LUSO BRASILEIRA SOBRE ACESSO ABERTO, 7., 2016, Viseu. **Proceedings...** Viseu: Portuguese Association of Librarians, Archivists and

Documentalists, 2016. n.2, p.71-79. Available at:
<www.bad.pt/publicacoes/index.php/cadernos/issue/download/71/pdf_19>. Accessed on: March 31, 2017.

SANTOS,N et al. **Knowledge management and the cognitive view of institutional repositories.**
Perspectivas em Ciencia da Informado, 2009.Available at:
<http://www.scielo.br/pdf/pci/v14n2/v14n2a15>. Accessed on: April 10, 2017.

SANTOS, H.M.D; FLORES, D. **Reliable digital repositories for archival documents:**
ponderations on long-term preservation. Perspectivas em Ciencia da Informado, v.20, n.2, p.198-218, abr./jun. 2015 Available at:
<Http://portaldeperiodicos.eci.ufmg.br/index.php/pci/article/view/2341/1604>. Accessed on: 07ago. 2017.

SAYAO, L. F; MARCONDES, C.H. Free software for institutional repositories: some subsidies for selection. In: SAYAO, Luis et al. (Org.). **Implementation and management of institutional repositories**: policies, memories, free access and preservation. Salvador: EDUFBA, 2009. p. 23-54. Available at:
<https://repositorio.ufba.br/ri/bitstream/ufba/473/3/implantacao_repositorio_web.pdf>. Accessed on: April 10, 2017.

SERRANO, A.P.S. **Information systems and quality**: the evaluation of archives and digital libraries. 2013. 253 f. Thesis (PhD in Information and Communication on Digital Platforms) - Faculty of Letters, University of Porto, Porto, 2013.

SILVA, A. M. da. **The impact of the widespread use of ICT (Information and Communication Technologies) on the concept of document** - Analytical-critical essay (II). Prisma.com, 2012, n.18. Available at: <http://revistas.ua.pt/index.php/prismacom/article/view/2229/pdf>. Accessed on: March 31, 2017.

SILVA, R.C; ADOLFO, L.B. **ARCHIVISTICS AND THE ARCHITECTURE OF INFORMATION:** AN INTERDISCIPLINARY ANALYSIS. Rio de Janeiro, 2006. Available at:
<http://www.brapci.ufpr.br/brapci/_repositorio/2009/11/pdf_39c36b8e5e_0006729.pdf>. Accessed on: August 18, 2017.

SILVA, D.F. **Reliable archival object repositories:** the roda software and perspectives for digital preservation. Porto Alegre, 2014. Available at:
<https://www.lume.ufrgs.br/bitstream/handle/10183/110741/000952831.pdf?sequence=1o>. Accessed on: October 28, 2017.

REDALYC SCIENTIFIC INFORMATION SYSTEM . **Red de Revistas Científicas de América Latina y el Caribe, España y Portugal.** 2015. Available at:
<http://www.redalyc.org/redalyc/media/redalyc_n/estaticasredalyc/acerca-de.html>. Accessed on: 12 Apr. 2017.

THOMAZ, K. **Trusted and Certifying Digital Repositories**. Rio de Janeiro, jan./jun.2007 Available at:<http://www.brapci.ufpr.br/brapci/_repositorio/2010/05/pdf_fed0720dbb_0010726.pdf>. Accessed on: March 29, 2017.

VECHIATO, F.L; FERNAL, A. **Digital repositories as archivists' working environments:** a study of archival principles and digital preservation in this context.
2013 .Available at: <http://www.uel.br/revistas/uel/index.php/infoprof/article/view/17272/pdf_1.>.

Accessed on: 28 Oct. 2017